Tamires Cruz Santos Silva
Antonio Martins

Elaboration and Characterisation of Mayonnaise Based on Moringa Oil

AF301669

Tamires Cruz Santos Silva
Antonio Martins

Elaboration and Characterisation of Mayonnaise Based on Moringa Oil

Comparison of oil content in Moringa seed by soxhlet extraction and LR-NMR analysis

ScienciaScripts

Imprint

Any brand names and product names mentioned in this book are subject to trademark, brand or patent protection and are trademarks or registered trademarks of their respective holders. The use of brand names, product names, common names, trade names, product descriptions etc. even without a particular marking in this work is in no way to be construed to mean that such names may be regarded as unrestricted in respect of trademark and brand protection legislation and could thus be used by anyone.

Cover image: www.ingimage.com

This book is a translation from the original published under ISBN 978-613-9-74321-6.

Publisher:
Sciencia Scripts
is a trademark of
Dodo Books Indian Ocean Ltd. and OmniScriptum S.R.L publishing group

120 High Road, East Finchley, London, N2 9ED, United Kingdom
Str. Armeneasca 28/1, office 1, Chisinau MD-2012, Republic of Moldova, Europe
Printed at: see last page
ISBN: 978-620-6-27914-3

Copyright © Tamires Cruz Santos Silva, Antonio Martins
Copyright © 2023 Dodo Books Indian Ocean Ltd. and OmniScriptum S.R.L publishing group

SUMMARY

CHAPTER 1		3
CHAPTER 2		4
CHAPTER 3		24
CHAPTER 4		29
CHAPTER 5		34
CHAPTER 6		35
CHAPTER 7		36

SUMMARY

Moringa oleifera Lamarck is a plant cultivated in many tropical countries, which has numerous popular uses due to its nutritional and pharmacological applications. This plant originates from northern India and belongs to the *Moringaceae* family. It has the characteristic of adapting to nutrient-poor soils and arid climates. Its seeds are of industrial importance as they produce an oil used to lubricate watches and other delicate machinery. It is also used in the manufacture of perfumes and in the chemical treatment of water. An interesting option to analyse the possibility of the seed oil being used in human food is by placing it as the base of one of the most consumed sauces in the world: mayonnaise. In this work, moringa oil was used to produce homemade mayonnaise, as a substitute for conventional oils. Physicochemical analyses of mayonnaise based on moringa oil were carried out, comparing it with mayonnaise based on soybean and sunflower oil, as well as oil characterisation through LR-NMR analysis. It is necessary to deepen the research on this, which can be a possible new product to be introduced in the market, if it is well accepted by the consumer.

Keywords: moringa, mayonnaise, LR-NMR, oils, fatty acids.

CHAPTER 1

INTRODUCTION

The development of new food products is an area that has been growing more and more because it provides consumers with a series of options when launching foods that are not conventional in certain niche markets, or inserting them in the formulation of products that are already well known by the population.

Moringa oleifera belongs to the Moringaceae family, which consists of only one genus (Moringa) and fourteen known species. Native to northern India, it now grows in several countries in the tropics. It is a small, fast-growing shrub or tree, reaching 12m in height. It has an open, umbrella-shaped crown and usually a single trunk. The flowers, which emerge in panicles, are cream-coloured, fragrant and much sought after by bees. The plant is known by several common names, according to its different uses. For some, it is known as 'drumstick' because of the shape of its fruits, which are a staple food in India and Africa. In some parts of West Africa, it is known as 'mother's best friend' as an indication that the local people know its full value. The plant produces a diversity of valuable products that local communities have made use of for hundreds, perhaps thousands of years.

The green fruits, leaves, flowers and roasted seeds are highly nutritious and consumed in many parts of the world. The oil obtained from Moringa seeds can be used in food preparation, soap making, cosmetics and as fuel for lamps.

An interesting option to analyse the possibility of the seed oil being used in human food is by placing it as the base of one of the most consumed sauces in the world: mayonnaise. In this work, our objective was to use moringa oil to produce homemade mayonnaise, as a substitute for conventional oils, to perform physicochemical analyses of mayonnaise based on moringa oil, comparing it with mayonnaise based on soybean and sunflower oil, as well as the characterisation of the oil through LR-NMR analysis.

CHAPTER 2

LITERATURE REVIEW

2.1 Moringa

Moringa oleifera Lamarck is a plant cultivated in many tropical countries, which has numerous popular uses due to its nutritional and pharmacological applications. This plant originates from northern India and belongs to the *Moringaceae* family. It has the characteristic of adapting to nutrient-poor soils and arid climates. It grows rapidly and can reach a height of 15 metres. Moringa leaves (Figure 1) are 25 to 45 centimetres long and elliptical in shape. The flowers (Figure 1) are white or beige in colour and 10 to 20 centimetres long. The fruit (Figure 1) is a kind of pod, which has two sides and a length of 20 to 60 centimetres. Inside are the seeds, always in large numbers, between 12 and 35 per pod, which have a diameter of approximately 1 centimetre and are winged (MOURA, 2009).

According to BECKER *et al.* (2001), moringa leaves are good sources of provitamin A, vitamins B and C, amino acids (methionine and cysteine) and minerals such as iron (582 mg/kg), potassium (21.7 mg/kg), calcium (26.4 mg/kg) and zinc (113.9 mg/kg). They are edible and usually eaten cooked like spinach and used in soups and salads. The green pods can be cooked and eaten as green beans.

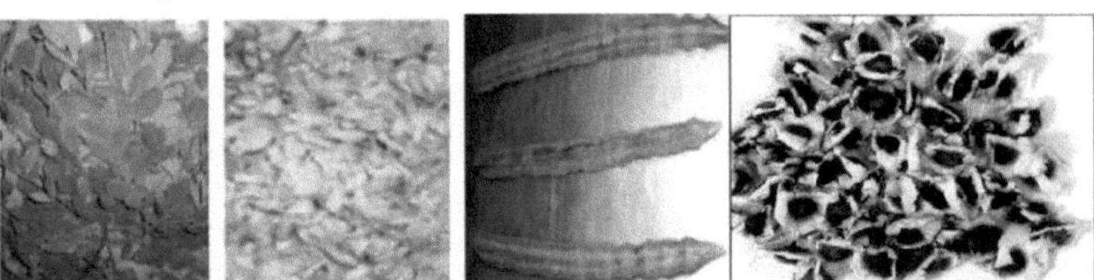

Figure 1: *Moringa oleifera* leaf, flower, pod and seed *in natura* Source: Moura, 2009.

Moringa flowers and leaves are sources of vitamins A, B, C and protein, and its dried seeds have a 30-42 per cent oil content.

The protein content of the moringa leaf does not vary considerably by region of cultivation. Studies carried out in Nicaragua, Bangladesh and India showed a protein content (dry basis) for the leaves of 25.1 %, 29.0 % and 26.4 %, respectively.

Moringa is a highly valued plant because it offers pods that can be consumed as preserves in vinegar, seeds that can be roasted, having a flavour similar to peanuts (GOPALAKRISHNAN *et al.,* 1980), as well as edible leaves and flowers that can be consumed in the form of sauces, salads or soups. The dried leaves can be used as a powder, partly replacing cassava flour.

The relative lack of anti-nutritional components and the high content of proteins, lipids, sulphur and amino acids, encourages the use of moringa for animal feed; it is an excellent source of protein for monogastric animals (OLIVEIRA et al., 2008). In addition to the benefits related to animal feed, Azevedo et al. (2018) found through experiments with diabetic rats that the aqueous extract of moringa leaves has a positive result for systemic and topical treatment of wounds in diabetic rats.

Dissemination programmes for the cultivation and use of moringa have been carried out in Brazil, given that the plant is an important source of provitamin A, with much higher levels than those found in vegetables such as broccoli, carrots, spinach and lettuce (CARVALHO *et al.,* 2008).

The main function of carotenoids is their ability to convert into vitamin A, whose functions in the body are related to vision, bone growth and tissue differentiation (IOM - U. S, 2001) & (OLSON & MACHLIN, 1991).

The role of carotenoids in the prevention and treatment of pathologies such as cancer, cardiovascular diseases, cataracts and prevention against degenerative disorders of the body has been reported in several studies (BENDICH, 1989; TAPIERO *et al.,* 2004; ZIEGLER, 1989).

Its seeds are of industrial importance as they produce an oil used to lubricate watches and other delicate machinery. It is also used in the manufacture of perfumes and in the chemical treatment of water (DUKE, 1987; MORTON, 1991). In Brazil, Moringa *oleifera has been* known in the state of Maranhão since 1950 (AMAYA et al., 1992). Currently, the culture of moringa has been disseminated throughout the semi-arid northeast, due to its use in water treatment for domestic use. The interest in the study of natural coagulants to clarify water is not a new idea. According to Ndabigengesere & Narasiah (1996), Moringa *Oleifera* seeds are a viable alternative coagulant agent to replace aluminium salts, which are used in water treatment worldwide. Compared to aluminium, *M. oleifera* seeds did not significantly alter the pH and alkalinity of the water after treatment and do not cause corrosion problems.

With the deepening concern with the preservation of the environment and the development of sustainable technologies, Lucena et al. (2016), carried out the study of green technologies that minimise the impacts on the environment without compromising the quality of the oil-derived pavement. Because Moringa *Oleifera* Lam seeds have an oil with antioxidant and lubricant properties, it was used to evaluate the changes in consistency properties and the variation of machining and compaction temperatures with the addition of it (extracted by pressing) as a green additive in six different proportions. The results showed that the addition of Moringa *Oleifera* Lam oil should be carried out at levels between 0.5% and 1%, since at these levels there is a significant decrease in machining and compaction temperatures (=5°C), without compromising the performance of the asphalt binder.

2.2 Mayonnaise

The consumption of industrialised foods and the interest of industries in the elaboration of these products have increased significantly in Brazil since the 70s. Among these foods, one of the best options is mayonnaise (SALGADO et al., 2006).

According to FURLANETTO, LACERDA, CERQUEIRA-CAMPOS (1982), mayonnaise is an emulsion composed of oil, eggs and vinegar, being considered a food with high lipid content. According to Brazilian legislation, this product must have a minimum of 65 g of edible vegetable oil/100 g of the product (BRASIL, 1978).

Mayonnaise has a pale yellow colour which derives from the egg yolks and not from the oil, although the chlorophyll in some oils can give a slightly greenish colour. The emulsion obtained in mayonnaise consists of two liquids which do not mix and which, after addition of emulsifiers and stirring, form a water-oil dispersion. The emulsion has an internal phase of oil droplets dispersed in an external phase of vinegar, eggs and other ingredients. The consistency presented by the emulsion depends on the ingredients used in the formulation, as well as the type of equipment and mode of operation during the manufacture of the product (PLUMER and GUIDOLIN, 2006).

The emulsion in mayonnaise is difficult to prepare. In it the major component, which is the oil, is forced to exist in the dispersed phase. The emulsion is more easily obtained when the fluid ingredients (oil and water) are cooled to a temperature of 12 to 15^0 C (PLUMER and GUIDOLIN, 2006).

Mayonnaise is a semi-perishable product, stable to be kept for a considerable time without refrigeration. In it we can find some types of alterations, so it has that the separation of the phases of mayonnaise is accelerated by the following factors: mechanical shock; exposure to very low temperatures; very rapid addition of oil in the preparation and unregulated agitation during emulsification (PLUMER and GUIDOLIN, 2006). The ingredients of the formulation composition are of paramount importance for obtaining an emulsion in the final product (PLUMER and GUIDOLIN, 2006):

Vegetable Oil. As mayonnaise is an oil-water emulsion, oil droplets are dispersed in an aqueous phase. The rigidity of the emulsion depends partly on the size of the oil droplets and how closely they are clustered. The more oil dispersed in the emulsion, the stiffer it is. The amount of oil added to mayonnaise is around 65 to 70%. When you reach 82% oil, there is an overload in the system and the droplets will be too close together and any mechanical check will easily cause the emulsion to degrade and the product will be much more fluid than creamy. Below 60% oil, the emulsion should be guaranteed by adding starch or by increasing the amount of egg yolk (PLUMER and GUIDOLIN, 2006).

Eggs. Egg yolk is the main emulsifying compound in mayonnaise formulation. The amount and type of egg solids have a pronounced effect on the viscosity and strength of the emulsion. It is the lecithoproteins, phospholipids and cholesterol in the egg yolk that bring the oil droplets and water droplets together and keep them stable, containing acetic acid (from vinegar), citric acid (from lemon juice), salt and sugar. Eggs are supplied in pasteurised (to eliminate salmonella) chilled or frozen form or as dehydrated powdered eggs. Frozen egg yolks provide a more uniform mayonnaise with a better consistency, while mayonnaise prepared with whole eggs is "thinner" and weaker because the white has less stabilising power than the yolk. Thus, the greater the amount of yolk, the greater the stability and viscosity of the mayonnaise. Dehydrated (powdered) egg yolks have a good emulsifying capacity, but provide a less stable emulsion than frozen egg yolks (PLUMER and GUIDOLIN, 2006).

Studies suggest that LDL-type lipoproteins adsorb the most at the oil-water interface and it is thought that these compounds, during adsorption, release phospholipids and proteins that adsorb at the interface and neutral lipids that coalesce with the oil droplets (HOCKERGÂRD, 2011).

Knowing that proteins exist at the emulsion interface, their stability depends largely on the pH and ionic strength of the medium. If these conditions are not favourable, the proteins will tend to denature, decreasing the repulsion forces between droplets (Hõckergârd, 2011). As egg yolk contains proteins with isoelectric points between five and eight, they are suitable for low pH emulsions, such as mayonnaise (HOCKERGÂRD, 2011).

An emulsifying agent, in order to be effective, must considerably reduce the surface tension between immiscible liquids; it must be fast to adsorb to the droplets of the dispersed phase, forming a film resistant to collisions between the droplets; it must have a structure with a polar zone orientated towards the aqueous phase and an apolar structure with affinity for the oil (figure 2); it must be more soluble in the aqueous phase, so that it is quickly available for adsorption and change the viscosity of the emulsion. From a commercial point of view, in addition to the above, it must be active at low concentrations and low cost (CHIRALT, 2005; MCCLEMENTSEWEISS, 2005).

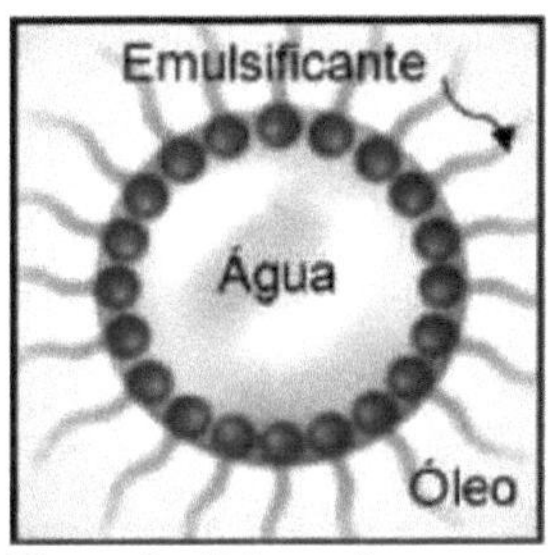

Figure 2 - Schematic representation of the interaction between oil, water and emulsifier (JAEGER, 2012).

Acids. Vinegar used is the main acid with a preservative function against microbiological spoilage of the product, although some

additives such as citric acid, sorbic acid and lactic acid can be used. Alcohol vinegar is the most commonly used and does not harm the colour of mayonnaise, which can happen if wine vinegar is used. The amount of vinegar added should provide a pH around 3.0 (PLUMER and GUIDOLIN, 2006).

Mustard. Mustard powder has emulsifying properties, as does the oil extracted from it, but it is used more for its organoleptic properties (spicy flavour) than for its emulsifying power. It has advantages and disadvantages both in powdered and essential oil form. Mustard powder contributes to the colour while essential oil does not. Mustard essential oil retains its pungency and flavour for a longer time and has no protein or starch, which can cause emulsion breakdown (PLUMER and GUIDOLIN, 2006).

Other Condiments. Salt and sugar are added at low levels. The salty or sweet flavour, however, is a function of their concentration in the aqueous phase of the emulsion. Since mayonnaise contains a higher amount of vegetable oil, the salt and sugar are dissolved in less water and therefore relatively concentrated. Other condiments can be used, depending on the product characteristics of each manufacturer (PLUMER and GUIDOLIN, 2006).

Food Additives. Food additives are substances that are added to foods in order to modify their organoleptic characteristics, nutritional value, processing techniques or conservation efficiency (REIS, 2013). According to

Freitas (2000), this type of substances have five central objectives, which are: maintaining the consistency of the product (emulsifiers give a consistent texture and prevent the food from disintegrating. Stabilisers and thickeners give uniformity and smoothness to the texture of the food); maintenance or improvement of nutritional value (addition of vitamins and minerals to common foods in order to complement nutritional deficiencies or replace some nutrients that may be lost during processing); maintenance of chemical and biological characteristics (preservatives minimise changes caused by microorganisms. Antioxidants prevent the development of rancidity and other undesirable oxidations); pH regulation (addition of acidifying or alkalising agents to modify the pH of a food and benefit its aroma, colour, shelf life, etc); aroma and colour control (some natural or synthetic spices and flavourings are used to alter or intensify the taste of food. Flavour enhancers are most commonly used in the food industry and have the function of increasing the intensity of flavour or adjusting a food to a certain type of taste that it did not originally have).

All food additives are subject to strict national and international legislation, in order to ensure, on the one hand, their harmlessness and, on the other hand, the correct labelling of foods. Thus, when a certain additive is approved for use in the food industry, regulations are published indicating the type of food in which it can be used, the maximum concentration allowed and how it should be mentioned on the labels (FREITAS and FIGUEIREDO, 2000).

One of the most important aspects when marketing mayonnaise is the preservation of the product over its shelf life. There are three key factors that can cause deterioration of mayonnaise, or mayonnaise-like products. These factors are emulsion instability, oxidation and hydrolysis reactions that can occur and lead to flavour deterioration, and microorganism growth that can lead to excessive acidification, also deteriorating the flavour of mayonnaise (VERMEULEN, 2008; DOORES, 2005). Thus, in the production of mayonnaise, the main additives used belong to the classes of preservatives, antioxidants, stabilisers, dyes and flavourings (REIS, 2013).

Lipids play an important role in the quality of certain food products, particularly in relation to the organoleptic properties that make them desirable (taste, odour, colour, texture). In addition to these, they confer nutritional value to foods, constituting a source of metabolic energy, essential fatty acids and fat-soluble vitamins (FRANKEL, 1996). In mayonnaise, fat is essential to flavour and texture and its reduction can affect the acceptability of the product (SALGADO et al., 2006).

Moreover, another factor that can affect the nutritional quality, safety, colour, flavour and texture of products rich in edible oils is the development of rancidity (FRANKEL, 1996 and SHAHIDI et al., 1992). This aspect is of great importance, not only from an economic point of view, due to losses from reduced shelf life, but also because of the possibility of the free radicals formed reacting or interacting with other constituents of the food, causing a drop in its nutritional quality (ANWAR, 1996).

Because recent research in the food area always seeks to meet new trends demanded by the consumer, investigating viable alternatives so that widely consumed products such as mayonnaise become increasingly rich from the nutritional point of view, without the loss of original organoleptic characteristics of the product, as well as increasing its shelf life, the substitution of soya oil, normally used to make mayonnaise, by moringa oil, becomes promising due to its high nutritional value, combined with an oxidative stability up to 60 times greater than that of soya oil, due to its high oleic acid content (ARAÚJO, 2006), of approximately 76%v (ANWAR, 2003), while soya oil has only 23% (MORETTO and FETT, 1998).

2.3 Oils

Oils and fats are water-insoluble substances (hydrophobic), of animal, vegetable or even microbial origin, formed predominantly from condensation products between "glycerol" and "fatty acids" called triglycerides (MORETTO and FETT, 1998).

The difference between oils (liquid) and fats (solid) at room temperature lies in the proportion of saturated and unsaturated acyl groups present in triglycerides, since the corresponding fatty acids represent more than 95% of the molecular weight of their triacylglycerols. Resolution No. 20/77 of the CNNPA (Council of Norms and Standards for Foods) defines the temperature of 20°C as the lower limit for the melting point of fats, classifying them as oil when the melting point is below this temperature. Animal fats, such as lard, edible tallow and butter, are made up of mixtures of triacylglycerols, which contain higher amounts of saturated acyl groups than unsaturated ones (MORETTO and FETT, 1998).

The word olive oil is used only for oils from fruits, such as olive oil and palm oil (MORETTO and FETT, 1998).

By definition, fatty acids are aliphatic monocarboxylic compounds derived from or contained in esterified form from a vegetable or animal fat, oil or wax, which commonly have a chain of 4 to 28 carbons, saturated or unsaturated (IUPAC, 1997).

The molecular mass of the glycerol (C_5H_5) portion of a triglyceride molecule is 41. The combined molecular mass of the fatty acid radicals (RCOO-) that make up the remainder of the molecule can range from about 650 to 970. Thus, fatty acids contribute about 94 to 96% of the total weight of the molecule. Because of their preponderant mass in the triglyceride structure, and their physical and chemical contribution to the properties of this molecule, fatty acids greatly influence the characteristics of glycerides (BAILEY'S INDUSTRIAL OILAND FATPRODUCTS, 1996).

With few exceptions, the fatty acids that occur in nature are straight-chain and have even numbers of carbon atoms. This idea that fatty acids were all even prevailed until 1954, when a branched-chain fatty acid, called "iso", was discovered in whale oil (BAILEY'S, 1979) and free fatty acids from human hair contained a carbon chain spanning from C7 to C22 (BAILEY'S, 1979) including fatty acids with carbon number C17 found in pork bacon fat (BAILEY'S 1979).

Fatty acids can be called saturated, when they do not have any double bonds, or unsaturated, in which case they may contain one or more double bonds. The degree of unsaturation depends on the number of double bonds the carbon chain can have.

After analysing the fatty acid composition of moringa seed oil (Figure 3), it was found that it has a percentage of oleic acid similar to the proportion present in olive oil (NORMAN, 1979), but with different proportions of the other fatty acids present (ANWAR et al., 2003). This fatty acid composition is also in agreement with what was reported by Ferrao (1970). Sengupta and Gupta (1970) also found similar result.

Figure 3. Moringa oil.
Source: Rural Sementes Ltda.

Due to this fatty acid composition (Table 1), moringa oil can be categorised as high oleic. High oleic oils are genetically difficult to reproduce, and have recently gained importance due to their superior degree of stability in addition to nutritional benefits (DELPLANQUE, 2000 and NEUZA et al., 1998).

tatty acid	M. oleifera[1]	literature
C"o	not detected	0.03 (30)
Cuo	not detected	0.11 (30)
C16O	6.50 + 0.40 (5.90-7.00)	6.04 (30)
CI6:1	1.00 ±0.08 (0.90-1.20)	0.11 (30)
0,8:0	5.67 ± 0.39 (4.88-7.50)	4.14 (30)
CIBzl	76.00 ±1.23 (74.05-78.59)	73.60 (30)
Col 8-2	1.29±0.10 (0.65-1.96)	0.73 (30)
CI8;3	not detected	0.22 (30)
C 200	3.00 ± 0.20 (2.95-4.21)	2.76 (30)
CzOl	1.20 ±0.06	2.40 (30)
C 22 0	5.00 ±0.33 (4.17-5.99)	6.73 (30)
Czzil	not detected	0.14 (30)
C26O	not detected	1.08 (30)

- Values are mean ± SD tor 12 *M obiera* oils analysed individually in triplicate. Parentheses show range ot data.

Table 1. Fatty acid composition of moringa oil.
Source: (Anwaretal., 2003 and Tsakinis et al., 1999)

The protein and oil contents in Moringa oleifera seeds are higher than those found in legumes and some varieties of soya beans, respectively.

However, there is some controversy regarding the content of anti-nutritional factors present in all parts of the plant. Anwar et al. (2003) states that further studies are needed on its nutritional potential, while Ferreira et al. (2008) states that the plant has low concentrations of anti-nutritional factors, although the seeds have glucosinolates (65.5pmol/g), phytates (41g/kg) and haemagglutinating activity, while the leaves have appreciable amounts of saponins (80g/kg), as well as phytates (21g/kg) and tannins (12g/kg). Also according to Ferreira et al. (2008), taking into account the excellent nutritional properties, the low toxicity of the seeds and the excellent ability of the plant to adapt to poor soils and arid climates, Moringa oleifera can be an alternative to the consumption of leguminous seeds, as a source of high quality proteins, oil and tannins.

antioxidant compounds.

2.4   Dispersions in food

An emulsion or a thermodynamically unstable dispersion of two immiscible liquids, usually of polar and apolar nature, in which one of them forms small-sized droplets (0.1 to 100 microns), which is called the dispersed or internal phase and the other, the continuous or external phase. In practice, this dispersion must contain a third important component: an emulsifier, whose nature of this substance is amphiphilic, which facilitates the formation of the emulsion. decreasing the interfacial tension between the apolar (oily) phase and the polar (aqueous) one, in addition to carrying a certain physical stability for a time, which can be longer or shorter, depending on the composition, processing characteristics and external conditions during storage of the product. Emulsions find applications in various fields: food, cosmetics, pharmacy, agricultural chemistry, detergents, paint and polymer industry, pre-treatment for oil in refineries, asphalts, etc (MUNOZ et al., 2007).

Almost all foods contain or are formed by dispersion of their components in water, forming single-phase systems (true solutions) or polyphase systems (suspensions). In true solutions, the solute particles have dimensions less than 10^{-6} cm, in suspensions, when the dispersed particles (dispersoids) have dimensions between 10^{-6} and 10^{-4} cm, they form colloidal dispersions; when the dimensions of the particles are above 10^{-4} cm, we will have suspensions (BOBBIO and BOBBIO, 1992).

In foods, colloids form the most important systems regulating the content and type of water present. The most frequent colloidal dispersions in foods are described in Table 2 (BOBBIO and BOBBIO, 1992):

Dispersed phase F.	Dispersant phase	$F, F,$	Classification	Examples
Solid	Liquid	S/L	Sun	Pectin solution
Liquid	Liquid	L/L	Emulsion	Mayonnaise
Gas	Liquid	G/L	Foam	Beaten egg white
Gas	Liquid	G/S	Solid foam	Ice cream

2. Types of colloidal dispersions in foods.
Source: (BOBBIO and BOBBIO, 1992)

Colloids of the L/L type, emulsions, are important because of their frequency in widely consumed foods, as shown in Table 3:

Food	Type of emulsion
Milk	0*/A*, stabilised by phospholipids and proteins.
Cream / Butter	AO, stabilised by phospholipids, proteins and synthetic emulsifying additives.
Margarine / Mayonnaise	A/0, stabilised by proteins, phospholipids and polysaccharides.
Cake batter	0/A, stabilised by proteins, phospholipids and polysaccharides
Ice cream / Mousse	0/A, stabilised by proteins, phospholipids and polysaccharides

0* = oil; A* = water

Table 3. Main types of emulsions in food.
Source: (BOBBIO and BOBBIO, 1992)

Stability of dispersions. The three main types of dispersions in food, suns, emulsions and foams, depend for their stability on several factors: density difference between the phases; viscosity of the system; dimensions of the dispersoid; ratio between volumes of the phases; existence and value of electric charges in the particles; presence of surface-active substances and surface tension at the interfaces. For each type of dispersion, one or more of the factors mentioned are of greater importance, alongside particular factors related to the nature of the phases (BOBBLO and BOBBIO, 1992).

The change in the ratio of the densities of the phases can lead to the separation of a phase by the coalescence of the particles of the dispersoid which, forming aggregates of increasing dimensions, may precipitate. The rate of precipitation will also depend on the viscosity of the medium which may hinder the movement of the dispersoid, slowing coalescence. The precipitation velocity is related to the density, viscosity and particle size by the following formula (BOBBIO and BOBBIO, 1992):

$$V = \frac{2}{9}r^2 \frac{(d - d_o)g}{\eta}$$

Where:

r = mean particle radius

of d_o = are the densities of the dispersoid and the dispersant

η = coefficient of viscosity of the dispersant

g = local gravity

The closer the densities of the phases are to each other, the easier it is to mix and homogenise them; the addition of thickeners is justified by the increased viscosity of the system. A low volume of the dispersed phase makes it difficult for the particles to approach, providing greater stability (BOBBIO and BOBBIO, 1992).

The main effect of the characteristics of the interfaces that must be considered in the formation of dispersions is the interfacial tension Y, which is the barrier to overcome in the homogenisation of the phases. The high interfacial tension of the components prevents their homogenisation, as it tends to resist the increase of the total contact surface between the phases (BOBBIO and BOBBIO, 1992).

The addition of substances that have lipophilic and hydrophilic groups (surfactants) in their molecules to a colloidal system produces a decrease in interfacial tension by their adsorption at the interface, with lipophilic and hydrophilic groups facing each phase respectively. With the decrease in surface tension, there can be an increase in the contact surface between the phases, allowing homogenisation of the system. Most emulsifying agents (surfactants) are also efficient foaming agents (BOBBIO and BOBBIO, 1992).

Emulsifiers. The choice of an emulsifier, or of mixtures, to obtain a stable dispersion, is based on the ratio that exists between its hydrophilic and lipophilic groups. When this ratio is expressed by the ratio between the percentages by weight of each group in the molecule, the value is called BHL, i.e.: hydrophilic-lipophilic balance (BOBBIO and BOBBIO, 1992).

Depending on the BHL value, the emulsifier will be lipophilic or hydrophilic and will be used in water/oil or oil/water emulsions. This empirical relationship is schematically represented in Table 4 (BOBBIO and BOBBIO, 1992):

BHL value	Character of the Emulsifier	Used in Emulsion
Less than 9	Lipophilic, poorly soluble in water	WaterOil
Between 9-11	Intermediate	
Above 11	Hydrophilic, very soluble in water	Oil/Water

Table 4: BHL values Source: (BOBBIO and BOBBIO, 1992).

BHL values can be calculated or determined experimentally. Their use in the choice of emulsifier is useful only to restrict the number of substances to be tried for each type of food.

2.5. Low Resolution Nuclear Magnetic Resonance (LR-NMR)

Nuclear Magnetic Resonance is one of the most important spectroscopic techniques in the study of biological systems, and in solid state physics and chemistry in general. Its importance stems from its successful application to a wide variety of problems, including technological ones, such as the design of high-resolution images of medical interest (H.R.I.) (SLICHTER, 1996).

This technique is an example of the interaction of matter with electromagnetic radiation, in which energy is absorbed or emitted according to the Bohr frequency condition $|AE| = hv$, where AE is the energy difference between the final and initial state of the matter under study, h is Planck's constant and v is the frequency of the radiation (GIL et al., 1987).

It is the nuclei of the atoms of the material sample that have a magnetic moment that intervene in this phenomenon: AE refers to nuclear magnetic states that have different energy when the sample is subjected to an external magnetic field. As the values of AE related to nuclear magnetic transitions are very small, the frequency v is within the typical values of radio frequencies (GIL et al., 1987).

The nucleus of an atom has a positive charge. When the spins of the protons and neutrons constituting the nucleus are not paired, the total spin of the nucleus generates a magnetic dipole around an axis. The intrinsic magnitude of this dipole is an important nuclear property called magnetic moment, p (FERREIRA, 2001):

$$\mu = \frac{\gamma I h}{2\pi}$$

Y is called the magnetogeometric ratio and is a constant for each nuclide; h is Planck's constant.

If a nucleus with a spin equal to 1/2 is placed under the influence of an external magnetic field, the magnetic moment of the nucleus will align both towards and against that magnetic field (FERREIRA, 2001).

The applied magnetic field produces an unfolding of the degenerate energy levels of the nuclear *spin*, so that transitions between them can be induced as a consequence of the absorption of an appropriate electromagnetic radiation (FERREIRA, 2001).

The energy of a given energy state is given by:

$$E = -\frac{\gamma h}{2\pi} m B$$

where B is the magnetic field strength of the core.

The energy difference between the levels (transition energy) is then represented by the expression (FERREIRA, 2001):

$$\Delta E = \frac{\gamma h B}{2\pi}$$

Imagine a nucleus (with *spin* equal to 1/2) in a magnetic field. This nucleus is at the lowest energy level (i.e. its magnetic moment does not oppose the applied magnetic field). The nucleus has a rotational motion around its axis.

In the presence of a magnetic field, this axis of rotation will perform a *precession* movement around the magnetic field (FERREIRA, 2001):

The precession frequency is called the *Larmor frequency and is* identical to the transition frequency and the potential energy of a nucleus that has undergone precession is given by (FERREIRA, 2001):

$$E = -\mu B\theta$$

where 0 is the angle between the direction of the applied magnetic field and the nuclear spin axis.

If energy is absorbed by the nucleus, then the precession angle, 0, will change. For a *spin 1/2* nucleus, the absorption of radiation "shakes" the magnetic moment so that it opposes the applied field (the higher energy state) (FERREIRA, 2001).

It is important to realise that only a small proportion of the nucleus is in the lower energy state and can absorb radiation. There is a possibility that when these nuclei are excited, the population of the higher and lower energy state becomes equal. Energy absorption is then no longer detected and *saturation of* the higher energy state is said to occur (SLICHTER, 1996).

There are *relaxation* mechanisms by which the nuclear *spin* system exchanges with the surrounding medium, allowing nuclei to return to the lowest energy state (FERREIRA, 2001).

Relaxation processes do not involve spontaneous radiation, since the probability of this occurring is very small for very close energy levels. They are transitions induced by fluctuating magnetic fields of appropriate frequency, which originate from the thermal agitation of the medium. The relaxation time is the rate at which each nucleus releases its energy and relaxes to the ground state. Energy can be expended by nuclei in two ways: - in a T_2 relaxation, also called spin-spin relaxation, in which a nucleus gives up its energy to another nucleus, in a process that causes no loss of energy in the surrounding nuclei; - in a T_i relaxation, or spin-net relaxation, in which a nucleus gives up its energy to its neighbourhood. For large amounts of energy, Ti relaxation leads to heating of the neighbourhood (CHANG, 1971).

The standard method of fatty acid determination based on gas chromatography, through methylation of the fatty acids present in the oil to be analysed (KURATA et al., 2005). is extremely time consuming compared to the LR-NMR method. The infrared spectroscopy method is another alternative for analysing intact samples, but it is slow and depends on the calibration of models from thousands of samples with similar chemical compositions. Conventional high-resolution nuclear magnetic resonance has been used for fatty acid analysis in oilseeds (COLNAGO et al, 1983 and FORATO et al, 2000), but the total time to obtain the results is too long for the demand of analyses required by genetic programmes (PRESTES et al., 2007).

The method using low resolution nuclear magnetic resonance has the potential to analyse more than 1000 samples per hour and is a powerful tool to select oilseeds with modified fatty acid profiles for food applications and biodiesel production (PRESTES et al., 2007).

The analysis of oilseeds is indispensable in the selection of grains to determine the energy value of food, biodiesel production and other uses. The NMR spectroscopic technique, one of the main analytical techniques, has been used to analyse the quantity and quality of oil in seeds (TOMA, 2009). According to Leal et al. (1981), the NMR method, besides being non-destructive, proved to be fast and useful for determining the types of oils from different seeds. The authors also recorded the 13C NMR spectra of a number of Brazilian oilseeds. The main constituents of the oils were identified and chemical shifts were estimated for the free fatty acids and the experimental shifts observed in the seeds and were correlated with literature data. Therefore, the authors concluded, as the number of reference spectra increases and certain steps are automated, the method could be employed for large-scale analysis of different species containing vegetable oils.

Rubel (1994) compared pulsed NMR techniques in analysing the amount of oil present in sunflower, linseed and soybean seeds. Routine analyses of oil and water content in different oilseeds were carried out by NMR because it is a rapid, accurate, precise and non-destructive technique. Pulsed NMR analyses of oil content and in mixtures showed equivalence and/or higher reproducibility than measurements made by traditional methods. Therefore, there were no statistical differences between the determination by the traditional method and by pulsed NMR. The author also observed that simultaneous determination of percentage of mixture and oil, present in seeds, was possible with pulsed NMR technique through spin-echo method. The addition of multiple components in oil can be detected and quantified by T2 analysis using CPMG pulse sequence (TOMA, 2009).

Hutton et al. (1999) used Magic- angle spinning (MAS) in 13C NMR spectroscopy, a practical and non-destructive method, for quantitative characterisation of oil composition in seeds. They described results for hybrid and intact canola seeds and the resulting biodiesel. According to the authors, the MAS 13C NMR results complement and agree with the results obtained by gas chromatography. The authors noted that MAS 13C NMR data allow quantitative analysis of the main components of the oil, including saturated and unsaturated fatty acids such as oleic, linoleic and linolenic. They concluded that this NMR method being non-destructive is attractive for plant breeding programme or other studies where loss of seed viability is inconvenient.

Azeredo et al. (2000) showed the use of equilibrium state free precession nuclear magnetic resonance (NMR/SSFP) for quantitative analysis at low magnetic field, exposing substantial advantages over conventional NMR methods. According to the authors, with only minor additional requirements, the technique allows a considerable increase in the signal-to-noise ratio for acquisition of a given time. The experimental conditions required for implementation and optimisation of parameter acquisition were explored and found to be easily accessible with unsophisticated equipment. The authors concluded that SSFP is a very useful technique for quantitative determinations performed on a low-field, low-resolution NMR spectrometer; and that because they are highly reproducible, results can be obtained quickly and without any dependence on standard sample.

Pedersen et al. (2000) analysed water, oil and protein content in rapeseed and mustard seeds by a combination of low-field 1H Nuclear Magnetic Resonance (LF-NMR) and chemometrics techniques, allowing the use of all relaxation curves in the evaluation of the data. Chemometric results were compared to traditional ones by Partial Least Squares Regression (PLSR) and Principal Component Analysis (PCA). The author stated that the classification of the two types of seeds was easily realised by LF-NMR (TOMA, 2009).

Weir et al. (2005) analysed the protein concentration in soybean seeds based on the amount of oil present in the seeds by high resolution NMR technique. The authors demonstrated that the selection of seeds with high protein content analysed by the amount of oil present in the seeds was reasonably effective in the population test and may offer a new rapid method for selecting soybean seeds with high protein content. According to the author, NMR technique has a potential to be used as an alternative analytical technique for indirect measurement of protein concentration based on amount of oil present in seeds (TOMA, 2009).

Terskikh et al. (2005) used the 13C NMR spectroscopic technique to analyse the amount of oil in seeds in vivo and the fatty acid composition in a non-destructive way.

Phippen et al. (2006) developed an efficient and reliable method to evaluate the oil content of *Cuphea* seeds and their respective fatty acids. Oil content was determined by the non-destructive NMR technique in all *Cuphea* seeds. According to the authors, the *Cuphea* species identified in this study may serve as potential new sources of oil and fatty acids to be introduced in the improvement of new strains (TOMA, 2009).

Colnago et al. (2007) developed a new method based on pulsed Nuclear Magnetic Resonance in Continuous Wave Free Precession (CWFP) steady state coupled to an on-line system (treadmill). The authors report that CWFP is a powerful technique for measuring oil content in intact seeds and has the potential to analyse more than 20,000 seeds per hour and is applicable to different seed species (TOMA, 2009).

Dais et al. (2007) evaluated analytical methodologies based on conventional 1H and 31P NMR spectroscopy and analytical methods (titration, gas chromatography, and high performance liquid chromatography) for measuring components (fatty acids and phenolic compounds), acidity and iodine index, in olive oil.

Prestes et al.(2007) demonstrated a rapid and automated 1H NMR method with low resolution, used to select intact oilseeds with different fatty acid contents, based on the transverse relaxation time constant (T2). T2 is measured through the decay of the NMR signal, obtained by CPMG and processed by chemometric methods, determining the quality of the oil in the seeds. Through its characteristics such as: fatty acid composition, cetane number, iodine index and kinematic viscosity.

CHAPTER 3

METHODOLOGY

3.1. Moringa Seed Oil Extraction and Extraction Yields

The oil extraction was performed by Soxhlet extraction system, which This type of extraction uses solvent reflux in an intermittent process, set up in the Natural Products Laboratory, Department of Chemistry, Federal University of Sergipe, based on method developed by ANDRADE et al. (2009).

The moringa seed portions (ripe, once and green), each averaging 29 g, were weighed and ground in an Arno multiprocessor, packed in a cartridge made with filter paper and placed in a Soxhlet extractor. Hexane was used, in a volume of 250 mL. After a period of 2 hours of extraction, the solvent was separated from the oil in a Fisatom rotary evaporator, from the Laboratory for the Analysis of Polluting Organic Compounds, also from the Chemistry Department of UFS.

After evaporation of the solvent, the oil was weighed on a Sartorius analytical balance, model TA215-S. The following formula was used to calculate the extraction yield:

$$Rendimento\ (\%) = \frac{Peso\ do\ balão\ com\ óleo - Peso\ do\ balão\ vazio}{Peso\ das\ sementes} \times 100$$

3.2. Analysis of Oil and Oleic Acid Content in Intact Moringa Seed by LR-NMR

Low resolution NMR experiments were performed on the SLK-SG-100 benchtop spectrometer (Spin Lock Magnetic Resonance Solutions, Córdoba, Argentina), operating at a resonance frequency of 9 MHz for the[1] H core, using the spin echo pulse sequence with QDamper pulses. The temperature of the samples was maintained at 25° C during the analyses, and each seed (ripe, once and green) was analysed individually in its oil and oleic acid contents, at the Magnetic Resonance Laboratory of Embrapa Instrumentação Agropecuária in São Carlos-SP. Data were acquired and processed using the SLK NMR Seed programme.

For the determination of water, total oil and oleic content in oilseeds by NMR-LR, the SLK NMR Seed software was developed by Spin Lock Magnetic Resonance Solutions, manufacturer of the SLK-SG-100 spectrometer (figure 4).

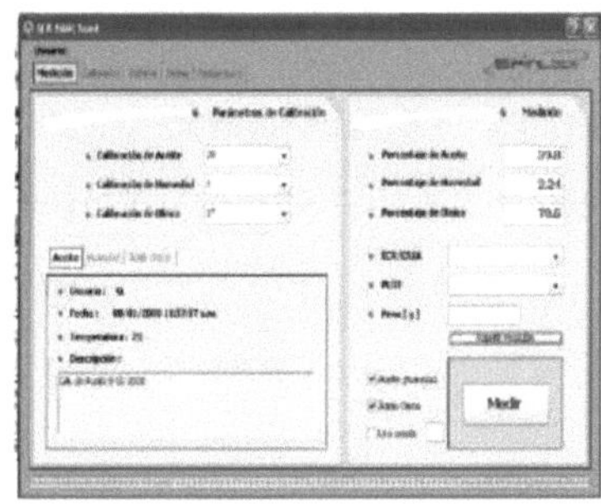

Figure 4 - Spin Lock NMR spectrometer, model SLK-SG-100 and SLK NMRSeed software.

Employing calibration curves, the software is able to determine oil and moisture content with a residual (difference between the observed value and the line estimated by the linear fit) of less than 1% and with repeatability (maximum allowable difference between measurements obtained on the same day) of less than 0.8% for 95% of measurements. The determination of oleic acid gives a residual of 5% and repeatability < 2% for 95% of measurements. The time for oil and moisture prediction is approximately 7 s and for oleic 30 s.

In addition to the specific programme, the SLK-SG-100 spectrometer has two advances over other low-field spectrometers, which make it possible to improve the quality of relaxometric measurements.

The determination of oil, moisture and oleic acid contents of oilseeds is based on calibration curves provided by the manufacturer, the so-called "standard curves". These curves are constructed from oilseeds with moisture, oil and oleic acid contents determined by standard methods, which are: drying to constant weight in an oven at 105°C, soxhlet extraction and gas

25

chromatography, respectively.

The NMR Seed programme allows the inclusion and use of user-produced calibration curves, enabling the analysis of other oilseed samples.

The NMR Seed V2 software uses the spin echo sequence with the addition of Qdamper pulses to determine the moisture and oil content on a wet basis (BU) and on a dry basis (BS) in oilseed samples. In its schematic representation (Figure 5), the value of s_i represents the sum of the intensities of the signals corresponding to oil and water and the value of S2 represents only the intensity of the oil signal. The difference between the two amplitudes (s_1-s_2) determines the intensity of the water signal.

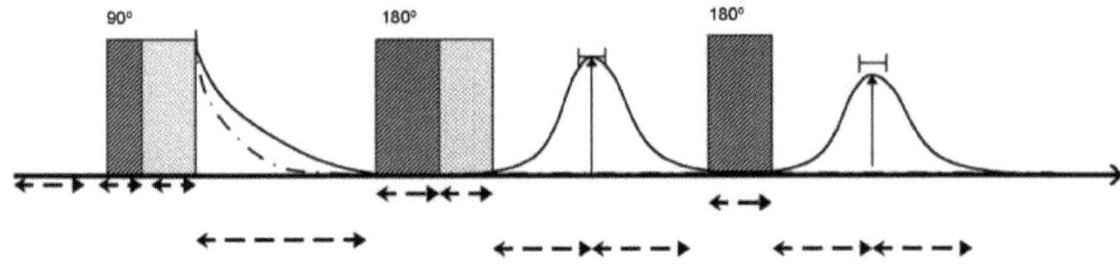

Figure 5 - Schematic representation of the spin echo sequence used for oil and moisture measurement in the NMR Seed software, where PW1 and PW2 represent the application time of the 90o and 180o rf pulses, respectively, QD the *Qdamper pulse,* Tau the echo time, Ad the echo acquisition time and TD *average* the time between cycles.

To determine the oil and moisture content, the values obtained by dividing the signal intensity of water (s_1-s_2) and oil (s_2) by the mass of the sample are correlated in a calibration curve. The calibration curve is generated by samples with moisture and oil content previously determined by standard techniques (in this case, soxhlet extraction and drying to constant weight, respectively).

The NMR Seed programme provides two oil values, oil on a wet basis (BU) and on a dry basis (BS). To calculate the BU oil content, the S intensity$_2$ is divided by the sample mass (I/m) and correlated with the corresponding oil content using the calibration curve, as mentioned above. For the calculation of the BS oil content, the mass corresponding to the percentage of water in the sample (which is calculated as water content predicted by s_2-s_1) is taken from the sample mass value. The value of S2 is then divided by this new mass value. In this way, the higher the moisture content of the sample, the higher the value of the BS oil relative to the BU oil.

To estimate the oleic acid content, the Seed NMR programme uses the CPMG (Carr-Purcel-Meigboon-Gill) pulse sequence with the addition of *Qdamper pulses* (Figure 6):

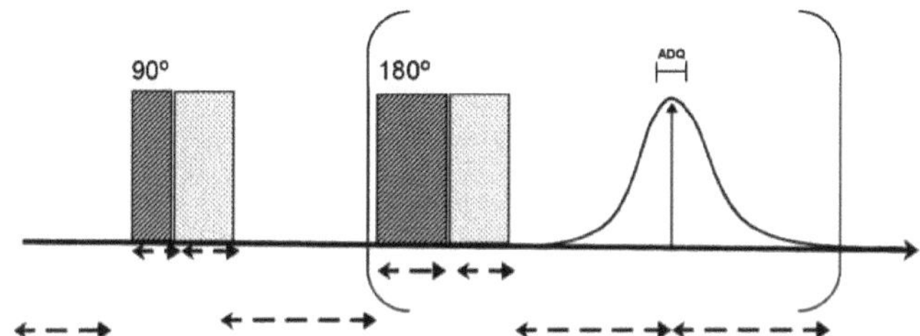

Figure 6 - Schematic representation of the CPMG pulse sequence with *Qdamper pulses used for the determination of oleic acid content in the NMR Seed programme.*
used to determine the oleic acid content in the NMR Seed programme, where PW1 and PW2 represent the time of application of the rf pulses of 90° and 180° , respectively, QD the Qdamper pulse, Tau the echo time, Ad the echo acquisition time, TD average the time between cycles and n the number of cycles.
Qdamper, Tau the echo time, Ad the echo acquisition time, TD *average* the time between cycles and n the number of echoes employed.

In oilseed samples, the value of T_2 depends on the composition and proportion of fatty acids present in the oil. Since, in general, viscosity increases with the length of the carbon chain and decreases with increasing unsaturation of the fatty acid with the same carbon number, the T_2 value can be used to analyse the quality of the oil in intact samples. It has been shown that CPMG signals coupled with chemometric methods are able to determine fatty acid composition, cetane number, iodine index and kinematic viscosity in seeds with good accuracy (Prestes *et al.*, 2007).

When the decay of CPMG data obtained in oilseed samples is fitted as biexponential, two T_2 values are obtained, the first in the order of 50 ms and the second in the order of. The shorter value corresponds to the oleic acid signal

(Pusiol *et al.,* 2003) and the longer one to the other fatty acids in the sample. As there is a large negative correlation between the longest T_2 value (of the order of 200 ms) and the determined oleic content, this value is used to predict the oleic content in the samples.

As temperature has a significant influence (Prestes, *et al.*, 2007) on the determination of T_2 time values, the value of the estimated oleic content may also change with temperature. It has been experimentally observed that there is a change in the order of 5% of the oleic content for each variation of 1°C. The manufacturer recommends keeping the sample at a temperature close to that of the equipment and waiting a few minutes inside the equipment before measuring the oleic content.

3.3. Preparation and Physical-Chemical Analyses of the Mayonnaise

Among the physicochemical analyses, determinations of ash, moisture, chlorides, titratable acidity and pH were performed, all according to A.O.A.C. (1995), in mayonnaises based on soybean (refined), sunflower (refined) and moringa (crude) oils, which were prepared based on the method of Bruscatto et al. (2004), with the following ingredients:

- 150mLoil
- 1 raw egg
- 1 egg yolk

- 3mL of white wine vinegar
- 1,72 g table salt
- 2g mustard-based sauce

The oil was beaten in a blender with the eggs for 15 minutes at high speed to form an emulsion. Soon after, the other ingredients mentioned above were added to the mixture, which were again mixed in a blender.

CHAPTER 4

RESULTS AND DISCUSSION

4.1 Moringa Seed Oil Extraction and Extraction Yields

As expected, mature seeds obtained maximum yield value, close to the yield range reached by Andrade et al. (2009), while "once" seeds obtained intermediate yield and green seeds, minimum yield, due to the fact that they had not yet reached their maximum maturation points:

- Mature seeds: 41.33%
- Seeds "for good": 31.94%
- Green seeds: 23.94%

4.2 Analysis of Oil Content in Intact Moringa Seed by LR- NMR

Low resolution NMR spectrometers operate with magnetic fields below two Teslas (80 MHz for[1] H) and provide data in the time domain.

With this technique, discrimination between the various components of a sample is based on differences in their transverse (T_2) and longitudinal (T_i) relaxation times. The determination of oil content in seeds is an example of this condition. The main constituents of oilseeds are carbohydrates, proteins, water and oil. The signals from solid materials, proteins, carbohydrates and bound water (water of hydration) have short T_2 (of a few microseconds) and decay rapidly. The NMR signals from free water and oil have long T_2 and take a few milliseconds to decay completely.

Using the spin echo pulse sequences (figure 7) the signal from the bound water and the other constituents of the sample is eliminated and only the oil signal is observed. The oil content of the sample is determined by means of a calibration curve, where the value of the signal intensity divided by the mass of the echo sample, leco/mass, is correlated on the calibration curve to obtain the corresponding oil content.

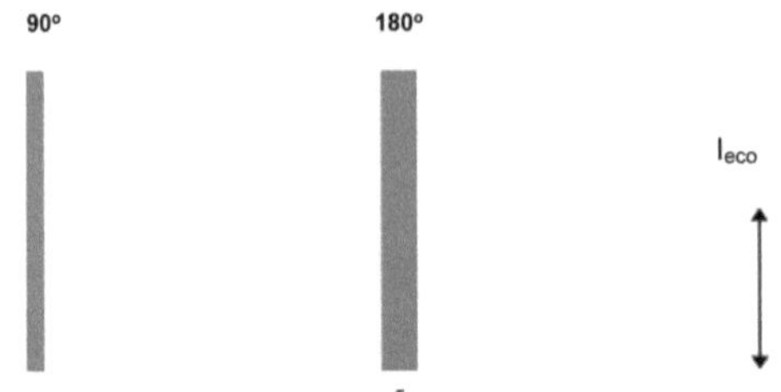

Figure 7. Spin echo sequence applied to the determination of oil in seeds. The dotted line
represents the signal from the solid material and water of hydration, while the black line
represents the signal from the oil.

After analysing the relaxation time of the samples, it was found that T_2 (1) (short) and T_2 (2) (long) of moringa resemble those found for jatropha and soya bean seeds as shown in Table 5:

Seeds	T_2 (1)(ms)	T2(2) (ms)
Moringa	46	151
Jatropha	43	147
Soya	43	144

Table 5. Comparison of transverse relaxation times of
moringa, jatropha and soya bean seeds.

Therefore, it can be inferred that calibration curves with jatropha and soya oils have similar viscosity to moringa oil which, in this case, is important for the replacement of the calibration curve. It was necessary to perform these measurements based on the calibration curve of other oils due to the current unavailability of moringa oil.

Curves made with oil of similar viscosity have similar angular coefficients and the results are then similar.

Below are the calibration curves of the jatropha and soya oils used:

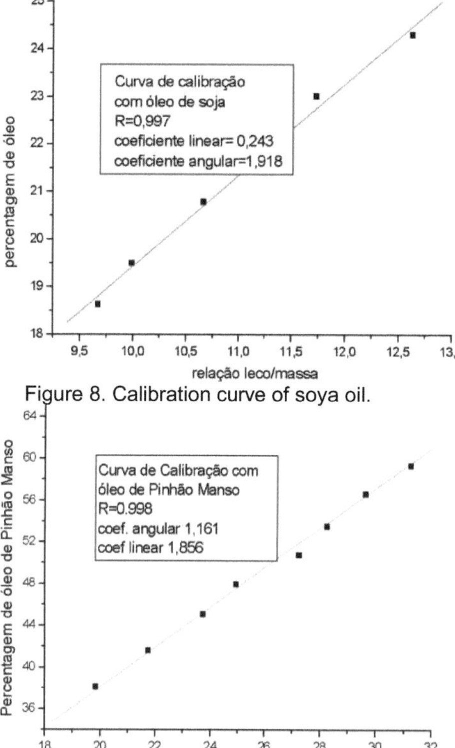

Figure 8. Calibration curve of soya oil.

Figure 9. Calibration curve of jatropha oil.

The intensity of the echo corresponds to the intensity of the NMR signal.

This value, divided by the mass of the oil, is related to the percentage of oil to construct the calibration curve.

Once the calibration curve is constructed, samples with unknown oil content are analysed. The NMR signal intensity value divided by the sample mass is then correlated on the calibration curve to obtain the oil percentage.

The oil content found in mature, "once" and green seeds were quite similar using NMR-LR. Because it has not yet been possible to construct a calibration curve for oil content and oleic acid content
with moringa oil, the latter was obtained with a calibration curve from sunflower seeds. This latter analysis is somewhat imprecise, due to the fact that it is very

variable with the temperature of the sample. The table below shows the oil and oleic acid contents found in the three types of seeds:

Seed maturation	Jatropha curve (% oil)	Soya Curve (% oil)	Average (% oil)	Sunflower curve (%oleic)
Green	42,9	43,4	43	88
"For good"	43,7	44,2	44	85
Mature	41,9	42,5	42	92

Table 6. Oil and oleic acid content of moringa seeds.

According to the results obtained, it can be stated that the seeds have a medium to high oleic content, in agreement with what was observed by Anwar et al., (2003), as a similar oil content to that reported by Andrade et al. (2009). However, it is worth mentioning that all these analyses need to be repeated with calibration curves constructed from the seeds and oil of moringa, in order to obtain more reliable results. The calibration curve for oleic acid should be constructed with samples whose oleic content has been determined by gas chromatography, which is the official method. Other samples with known oleic content can then be used to analyse the reliability of the constructed curve.

1.3. Preparation and Physical-Chemical Analyses of Mayonnaise

Regarding the preparation of the mayonnaise, both the mayonnaise made from soya oil and the one made from sunflower oil obtained organoleptic characteristics similar to the mayonnaise available for sale on the market. On the other hand, the mayonnaise prepared from moringa oil acquired a more yellowish colour, as well as a strong unpleasant odour and somewhat different from that obtained with the preparation of the other mayonnaises, in addition to the formation of bubbles. It is possible that these problems are made more evident by the fact that the oil was used in its crude form, which means that a number of substances are present in its composition (free fatty acids, phospholipids, metals, chlorophyll and oxidation products) which, above certain limits, can instabilise the emulsion of the product. These undesirable substances are totally or partially removed during the oil refining process. The unpleasant odour may also have been caused by some hexane residue present in the crude oil.

Below are the results obtained from the physical-chemical analyses carried out on the three types of mayonnaise:

Mayonnaise	Acidity (%)	PH	Ash (%)	Moisture (%)	Chlorides (%)
MOM[1]	3,40	5,09	0,24	35,95	0,91
MOS[2]	1,58	5,80	0,96	76,61	4,48
MOG[3]	1,52	5,82	1,24	73,99	6,22

[1]Moringa oil-based mayonnaise
[2]Mayonnaise based on soya oil
[3]Sunflower oil-based mayonnaise

Table 7. Results of the physicochemical analyses of the mayonnaises

The fact that the acidity percentage of MOM is almost double in

The lower pH of MOM compared to MOS and MOG may be due to the fact that crude moringa oil contains a much higher proportion of free fatty acids in its composition when compared to other oils that are refined and also to the fact that, during the preparation of MOM, it received a double amount of white wine vinegar in its formulation, compared to MOS and MOG. The lower pH of MOM than MOS and MOG also attests to this hypothesis.

The percentage of ash in MOM (highlighted in red) cannot serve as a comparison value with the other mayonnaises, because the oil used for the preparation of MOM contains a large amount of solid particles in suspension (a higher ash content would be expected than in MOS and MOG), and the relative deviation between the ash determinations carried out exceeded 100%.

The lower moisture value found in MOM can also be explained by the high content of suspended solid particles in the moringa oil.It is possible that the chloride content of MOM has such a small value (5 to 7 times lower) compared to MOS and MOG due to the non-addition of salt during its preparation. The higher chloride content found in MOG may be due to a higher quantity of mustard-based sauce added to it.

CHAPTER 5

CONCLUSIONS

From the results obtained in this work, it is clear that further investigation of the characteristics related to moringa oil-based mayonnaise is necessary, through the repetition of all the analyses carried out, accompanied by other determinations and procedures (characterisation and refining of the oil, complete centesimal composition, toxicological analyses, microbiological analyses, etc.), so that it can be ascertained whether there is a possibility that this product has a promising insertion in the food market.

CHAPTER 6

FUTURE PERSPECTIVES

> To carry out refining and physicochemical characterisation of moringa oil, as well as fatty acid analysis by GC-MS;

> To check if the anti-nutritional factors of the seed remain in the oil even after extraction and refining;

> Repeat the analyses performed by NMR-LR, with the construction of calibration curves for moringa oil;

> Complete centesimal composition of the mayonnaise, accompanied by microbiological analyses, if moringa oil is found not to be toxic;

> If all the previous steps are successfully completed, determine shelf life, perform ADQ and consumer acceptance testing and, based on the results of the sensory analyses, seek formulation improvements for scale-up.

CHAPTER 7

BIBLIOGRAPHICAL REFERENCES

AMAYA, D. R.; KERR,W. E.; GODOI, H. T.; OLIVEIRA, A. L.; SILVA, F. R. Moringa: arboreal vegetable rich in beta-carotene. Horticultura Brasileira, v.10, n.2, p.126, 1992.

ANDRADE, A.G.; JÚNIOR, L.A.R.; SANTOS, R. B.; SOLETTI, J. I.; CARVALHO, S. H. V. Study of the extraction process of moringa oleifera oil. National Meeting of Moringa, 02 to 04 September 2009. Aracaju - Sergipe.

ANWAR, F.; BHANGER, M. I. Analytical Characterisation of *Moringa oleifera* Seed Oil Grown in Temperate Regions of Pakistan. *J. Agric. Food Chem.* 2003, *51,* 6558-6563.

AOAC INTERNATIONAL. Official Methods of Analyses of AOAC, 16 ed. Arlington (USA), 1995.

ARAÚJO, J. M. A. Química de alimentos: teoria e prática /Viçosa; UFV; 20042006. 478 p.

AZEREDO, R. B. V.; ENGELSBERG, M.; COLNAGO, L. A. Quantitative analyses using steady-state free precession nuclear magnetic resonance. Analytical Chemistry, v. 72, n.11,p. 2401-2415, 2000.

AZEVEDO, I.M.; ARAUJO-FILHO, I.; TEIXEIRA, M.A.A.; MOREIRA, M.D.F. de C.; MEDEIROS, A.C. Wound healing of diabetic rats treated with Moringa oleifera extract. Acta Cir Bras. 2018;33(9):799-805.

AZOUBEL, L.M.O *et al.* Food composition table. *In*: DUTRA-DE- OLIVEIRA, J.E., MARCHINI, J.S. *Ciências nutricionais*. São Paulo : Sarvier, 1998. p.363-376.

BAILEY'S INDUSTRIAL OIL AND FAT PRODUCTS 4.ed.NY: 1979.Vol 1, pg 312.

BAILLEY'S INDUSTRIAL OIL & FAT PRODUCT, 5.ed. NY:1996. V1, Chap 11.

BECKER, K. FOIDL N., MAKKAR H.P.S. "The potential of *Moringa oleifera* for agricultural and industrial uses", Dar Es Salaam, October20th, November 2nd, 2001.

BENDICH, A. "Carotenoids and the immune response". J. Nutr., v. 119, p. 112115, 1989.

BOBBIO, P. A.; BOBBIO, F. O. Química do processamento de alimentos. 2 ed. São Paulo: Varela, 1992.

BRAZIL. Official Gazette of the Union. Laws, decrees, etc. Ordinance 12/78 of the CNNPA. Brasilia, 1978.

BRUSCATTO, M. H.; CARBONERA, N.; CONCEIÇÃO, Q. A.; DREWS, C.; TEIXEIRA, A.M.; ZAMBIAZI, R. Elaboration of mayonnaise with different types of vegetable oils. 2004. Available at: www.ufpel.edu.br/cic/2004/arquivos/CA00832.rtf

CARVALHO, A.F.U. *Moringa oleifera:* bioactive compounds and nutritional potentiality, Revista de nutrição - Campinas, SP, 2007.
CHANG, R. Basic Principles of Spectroscopy, Ed. AC, Madrid (1971).

CHIRALT, A. (2009). Food Emulsions. In Barbosa-Canovas, G.V., Food Engineering,. 1, 1-39.

COLNAGO, L.A. ; SEIDL, P.R. J. Agric. Food Chem. 31 (1983) 459.

COLNAGO, L. A. et al. High-throughput, non-destructive determination of oil contentin intact seeds by continuous wave-free precession NMR. Analytical Chemistry, v. 79, n. 3, p. 1271-1274, 2007.

COSTA, P. A. da. Characterisation of fatty acids, tocopherols and phytosterols in fruits and nuts from the north and northeast regions of Brazil. Dissertation (Master). Scielo Brasil, vol. 21, n^0 4, Campinas Jul/Aug, 2008.

DAIS, P.; SPYROS, A. 31P NMR spectroscopy in the quality control and authentication of extra-virgin olive oil: a review of recent progress. Magnetic Resonance in Chemistry, v. 45, n. 5, p. 367-377, 2007.

DELPLANQUE, B. The nutritional interests in sunflower oil: sunflower seed with linoleic and high oleic acid contents. *Ol., Corps Gras, Lipides* 2000, 7 (6), 467-472.

DOORES, S. Organic acids. In: DAVIDSON, P.M., SOFOS, J. N., AND BRANEN, A. L., Antimicrobials in food. CRC Press, Boca Raton, 91-142, 2005.

DUKE, J. A. Moringaceae: horseradish-tree, drumstick-tree, sohnja, moringa, murunga-kai, mulungay. In: BENGE, M. D. (Ed.) Moringa a multipurpose tree that purifies water. Boston, Science and Technology for Environment and Natural Resources, 1987, p.19-28.

FERRAO, A. M. B.; FERRAO, J. E. M. Fatty acids in moringa. *Agron. Angolana* 1970, *30*, 3-16.

FERREIRA, C. S. de O. Molecular Modelling and Nuclear Magnetic Resonance Studies of Inclusion Complexes between Cyclodextrins and Gadolinium (III) polyazamacrocycles. Dissertation (Master). Faculty of

Sciences of the University of Porto, Portugal, 2001.

FERREIRA, P. M. P.; FARIAS, D.F.; OLIVEIRA, J. T. de A.; CARVALHO, A. de F. U. Moringa oleifera: bioactive compounds and nutritional potential. Rev. Nutr., Campinas, 21(4):431-437, Jul./Aug., 2008.

FORATO, L.A.; COLNAGO, L.A.; GARRATT, R.C.; LOPES, M.A. Biochim. Biophys. Acta 1543 (2000) 106.

FRANKEL, E. N. Antioxidants in lipid foods and their impact on food quality. Food Chem., Kidlington, v. 57, n.1,p. 51-55, 1996.

FREITAS, A.; FIGUEIREDO, P. Food Conservation. Lisbon, 2000.

FURLANETTO, S. M. P.; LACERDA, A. A.; CERQUEIRA-CAMPOS, M. L. Research of some microorganisms in salads with mayonnaise purchased in restaurants, snack bars and "rotisseries". Rev. Saúde Pública, São Paulo, v. 16, n. 6, 1982.

GIL, V. M. S.; GERALDES, C. F. G. C. *Nuclear Magnetic Resonance: Fundamentals and Applications.* Calouste Gulbenkian Foundation, Lisbon (1987).

GOPALAKRISHNAN, P.K. *et. al.* Drumstick (*Moringa oleifera*) A Multipurpose Indian Vegetable, Economy Botany, 34 (3), pp 276 - 283, 1980.

HOCKERGÂRD, A. (2011). The Freeze-Thaw Stability of Mayonnaise and the Effect of Octenyl Succinic Anhydride Modified Starch as Emulsifier. Linn^us University in Kalmar, Kalmar.

HUTTON, W. C.; GARBOW, J. R.; HAYES, T. R. Nondestructive NMR determination of oil composition in transformed canola seeds. Lipids, v. 34,

n. 12, p. 1339-1346, 1999.

IOM - U. S. Institute of Medicine, Food and Nutrition Board, Standing Committee on the Scientific Evaluation of Dietary Reference Intakes. "Dietary Reference Intakes: for Vitamin A, Vitamin K, Arsenic, Boron, Cromium, Copper, Iodine, Iron, Manganese, Molybdenium, Nickel, Silicon, Vana-dium and Zinc". Washington, D.C., National Academy Press, 797 p. 2001.

IUPAC, (1989) The nomeclature of steroids. Retrieved from: http://www.chem.qmul.ac.uk/iupac/steroid/

JAEGER, J. (2012). Mayonnaise production. Work presented for evaluation in the discipline of Planning and Projects of Industry II of the Chemical Engineering Course of the Centre of Technological Sciences of the Regional University of Blumenau. Regional University of Blumenau, 160 p.

KURATA, S.; YAMAGUCHI, K.; NAGAI, M. Anal. Sci. 21 (2005) 1457.

LEAL, K. Z. et al. Immediate analysis of the oleaginous content of seeds by carbon-13 nuclear magnetic resonance. Ciência e Cultura, v. 33, n. 11, p. 1475-1484, 1981.

MARCONCINI, L. V.; SANO, S. M.; COLNAGO, L. A. Determination of Oil Content in Baru Seeds by Low Resolution NMR. III Congress of the Brazilian Biodiesel Technology Network - RBTB, 09 and 10 November 2009. Brasília - Federal District.

MCCLEMENTS, D.; WEISS, J. (2005). Lipid Emulsions. in: Shahidi, F. (Eds.) Bailey's Industrial Oil and Fat Products. 6ª Edition, John Wiley & Sons, Massachusetts, USA, 457-502 p.

MORTON, J. The horseradish tree, *Moringa pterygosperma* (Moringaceae) - a boon to arid lands? Economy Botany, v.45, n.3, p.318-333, 1991.

MOURA, A. S.; SOUZA, A. L. G.; OLIVEIRA JÚNIOR, A. M.; LIRA, M. L.;

SILVA, G. F. Physicochemical Characterisation of the Leaf, Flower and Pod of Moringa *(Moringa oleifera* Lamarck). National Moringa Meeting, Aracaju, Sergipe, 2009.

MUNOZ, J.; ALFARO, M. del C.; ZAPATA, I. Avances en la formulación de emulsiones. GrasasyAceites, 58 (1), enero-marzo, 64-73, 2007.

NAWAR, W. W. Lipids. In: FENNEMA, O. R. Food chemistry. 3.ed. NewYork: Marcel Dekker, 1996. p. 225-319. (Food science and technology).

NDABIGENGESERE A.; NARASIAH, S. K.. Influence of operating parameters on turbidity removal by coagulation with Moringa oleifera seeds. Environmental Technology, v.17, p.1103-1112, 1996.

NEUZA, J.; LIRENY, G. Behaviour of high oleic sunflower heated under thermal oxidation and frying conditions. *Cienc. Tecnol.* 1998, *18*(3), 335-342.

NORMAN, O. V. S. Composition and characteristics of individual fats and oils. In *BaileyS Industrial Oil and Fat Products*, 4th ed.; Swern, D., Ed.; Wiley: NewYork, 1979; Vol. 1, pp 289-459.

OLIVEIRA, L. H. L C. de; BASSETTO, M. N. Product: Mayonnaise. Instituto de QuímicadaUnicamp . Disponívelem : http://pcserver.iqm.unicamp.br/~wloh/offline/qg661/work14.html

OLIVEIRA, I.C.; TEIXEIRA, E.M.B.; GONÇALVES, C.A.A.; PEREIRA, L.A. Centesimal Evaluation of Moringa Oleifera Lam Seed. II Scientific Initiation Seminar - IFTM, Campus Uberaba, MG. 20 October 2009.

OLSON, J.A. In: MACHLIN, L.J., ed. Handbook of Vitamins. 2ª ed. New York, Mareei Dekker, p. 1-57. 1991.

PALADA, M. C. Moringa (Moringa oieifera Lam.): a versatiie tree erop with hortieuiturai potentiai in the Subtropieai United States. Hortieuiture Seienee, v.31, n.5, p.233-234, 1996.

PEDERSEN, H. T.; MUNCK, L.; ENGELSEN, S. B. Low-eid 1H nueiear magnetie resonanee and ehemometries eombined for simuitaneous determination of water, oii, and protein eontents in oiiseeds. Journai of the Ameriean Oii Chemists' Soeiety, v. 77, n. 10, p. 1069-1076, 2000.

PHIPPEN, W. B.; ISBELL, T.; PHIPPEN, M. E. Totai seed oii and fatty aeid methyi ester eontents of euphea aeeessions. Industriai Crops and Produets, v. 24, n.1,p. 52D59, 2006.

PIGHINELLI, A.L.M.T.; PARK, K. J.; RAUEN, A. M.; ANTONIASSI, R. Optimisation of the Transesterifieation of Girassoi Oieo. Biodiesei: o novo eombustívei do Brasii. Faeuidade de Engenharia Agríeoia, Unieamp, Campinas, SP.

PINHEIRO, A.B.V., Tabeia para avaiiação de eonsumo aiimentar em medidas easeiras. 4.ed. São Pauio : Atheneu, 2001.

PLUMER, E. da C.; GUIDOLIN, F. Fabrieação de Maionese. Brazilian Service of Answers Téenieas. SENAI-RS, 2006.

PRESTES, R. A., COLNAGO, L. A., FORATO, L. A., VIZZOTTO L., NOVOTNY, E. H., CARRILHO E. A rapid and automated iow resoiution NMR method to anaiyze oii quaiity in intaet oiiseeds, Anaiytiea Chimiea Aeta, 596, 325, 2007.

PUSIOL, D. J., ZURIAGE, J. J. Spin-Spin and Spin Lattice relaxometry in sunflowers seeds, G. in: Magnetic Resonance in Food Science, Lasted Development, Edited by Belton, P.S., Gil, A.M. Webb G.A. and Rutledge D. A.,

Royal Society of Chemistry, UK, Food Trade Press Ltd, pg 93, 2003.

REIS, J.P.M.F. Development of New Formulations of Traditional, Light and Fat-Free Mayonnaise. Dissertation (Master). Faculty of Science and Technology. Universidade Nova de Lisboa. Portugal, 2013.

RUBEL, G. Simultaneous determination of oil and water contents in dierent oilseeds by pulsed nuclear magnetic resonance. Journal of the American Oil Chemists' Society, v. 71, n. 10, p. 1057-1062, 1994.

SALGADO, J. M.; CARRER, J. C.; DANIELI, F. Sensory evaluation of traditional mayonnaise and mayonnaise enriched with aromatic herbs. Ciênc. Tecnol. Aliment., Campinas, 26(4): 731-734, Oct.-Dec. 2006.

SENGUPTA, A.; GUPTA, M. P. Studies on seed fat composition of Moringaceaefamily. *Fette, Seifen, Anstrichm.* 1970, 72(1), 6-10.

SHAHIDI, F.; WANASUNDARA, P. K. J. P. D. Phenolic antioxidants. Crit. Rev. Food Sci. Nutr., Lauderdale, v. 32, n.1,p. 67-103, 1992.

SLICHTER, C.P. *Principies of Magnetic Resonance,* Springer Verlag, Berlin, (1996).

TAPIERO, H.; TOWNSEND, D.M.; TEW, K.D. "The role of carotenoids in the prevention of human pathologies". Biomedicine & Pharmacotherapy, v. 58, p. 100-110, 2004.

TERSKIKH, V. et al. In vivo 13C NMR metabolite profiling: potential for

understanding and assessing conifer seed quality. Journal of Experimental Botany, v. 56, n. 418, p. 2253D2265, 2005.

TOMA, D. Quality analysis of vegetable oils in intact seeds by low resolution NMR. Dissertation (Master). Chemistry Institute of São Carlos, University of São Paulo. São Carlos, 2009.

TSAKNIS, J.; LALAS, S.; GERGIS, V.; DOURTOGLOU, V.; SPILITOIS, V. Characterisation of *Moringa oleifera* variety Mbololo seed oil of Kenya. *J. Agric. Food Chem.* 1999, *47*, 4495-4499.

VERMEULEN, A. Microbial stability and safety of acid sauces and mayonnaise-based salads assessed through probabilistic growth/no growth models. Ghent: Faculty of Bioscience Engineering, University of Ghent, 2008.

WEHRLI, F. W.; SHAW, D.; KNEELAND, J. B. *Biomedical Magnetic Resonance Imaging, Principies Methodology and Applications,* VCH Publishers NewYork (1988).

WEIR, A. D. et al. Use of NMR for predicting protein concentration in soybean seeds based on oil measurements. Journal of the American Oil Chemists' Society, v. 82, n. 2, p. 87-91, 2005.

ZIEGLER, R.G., "A review of epidemiologic evidence that carotenoids reduce the riskf cancer". J. Nutr., v. 119, p. 116-122, 1989.

I want morebooks!

Buy your books fast and straightforward online - at one of world's fastest growing online book stores! Environmentally sound due to Print-on-Demand technologies.

Buy your books online at
www.morebooks.shop

Kaufen Sie Ihre Bücher schnell und unkompliziert online – auf einer der am schnellsten wachsenden Buchhandelsplattformen weltweit! Dank Print-On-Demand umwelt- und ressourcenschonend produziert.

Bücher schneller online kaufen
www.morebooks.shop

MIX
Papier aus verantwortungsvollen Quellen
Paper from responsible sources
FSC® C105338

FSC
www.fsc.org

Printed by Books on Demand GmbH, Norderstedt / Germany